Introduction

This book discovers child's natural tendency to learn Mathematics.

It doesn't ask the readers to do any thing special but it asks to notice what child does and bring that into their notice.

Readers of all age group will enjoy this book but it is designed for the age group 0 to 5.

While going through this book you will notice that examples and pictures are used, are mostly from Nature.

<u>Let's play with Maths</u> has two volumes; the present book is volume one of this series, which introduces the basic concepts of maths to the tiny brains.

Author

Sunil kushwaha

Dedication

This book I would like to dedicate to my son Devendra and Daughter Hritika .

INDEX

Chapter

1. Counting in the Lap of Nature..................04

2. Shapes in the Surroundings.....................10

3. A point..26

4. A line segment....................................29

5. A Ray...33

6. A Line 37

7. A Plane..40

Let's check your understanding..................43

Let's see Where We use Maths...................47

What does Volume-2 consist.....................52

Lesson 01   Counting in the Lap of Nature

Take your kids and go to sea beaches. Ask them to see the nature.

Nature is a great teacher. Nature itself is a great book. Nature has unlimited elements to teach to your kids and to you.

Ask them to the see the birds, animals, trees, and sea waves.

Ask them to count the number of boats?

Inspire them to count the birds flying in the sky.

Do you see that the lake works as a mirror, this is known as reflection.

———————

Lesson 02 Shapes in the Surroundings

Our nature is full of different geometrical Figures. A straight (line) segment of bridge, (apparent) circular sun,

Leaves of different shapes, stones, each and everything in the universe has its unique shape.

While we are teaching to our kids we must take aid of things available in the surrounding. A child cannot carry books to everywhere but it is Nature which he finds everywhere.

Identify the shapes of following things:

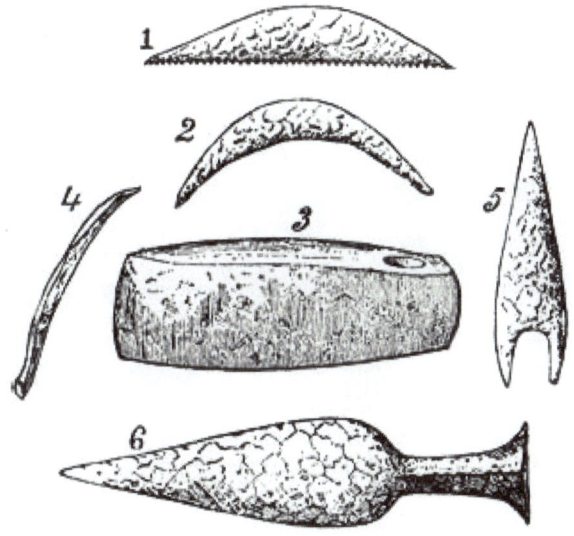

In above picture one may find difficult to identify Shapes because they are combination of different Shapes .Here your kid is not expected to answer correctly, but our goal is to promote thinking and create interest for Maths.

Sphere

Rectangle:

Cover page of a book

Cuboid:

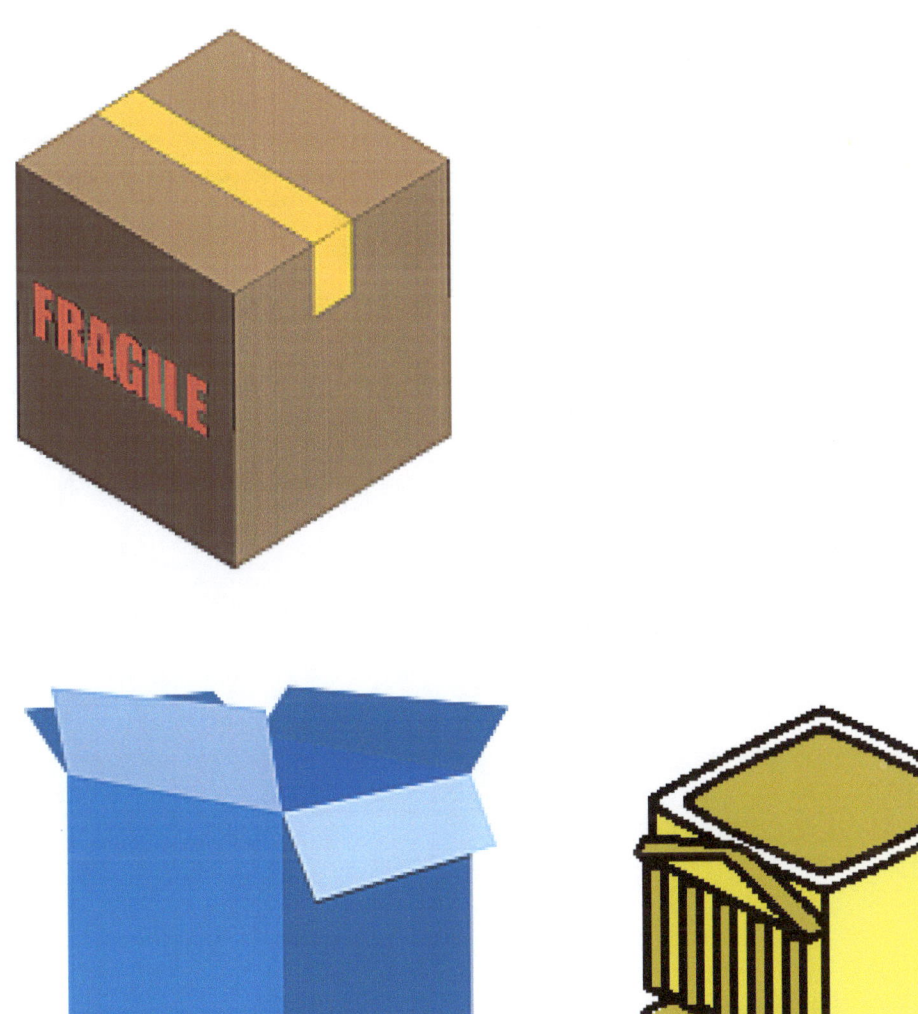

Cylinder:

Circle:

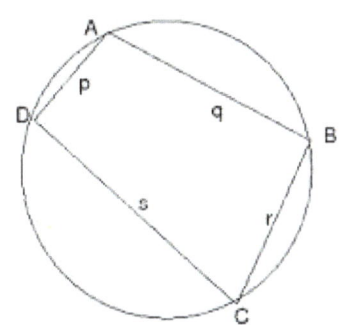

Square:

Triangle

Cone:

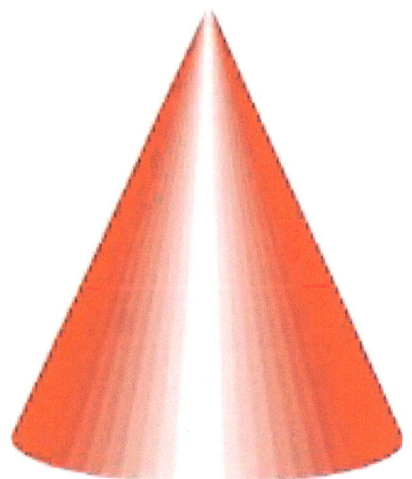

Frustum:

Semi-circle:

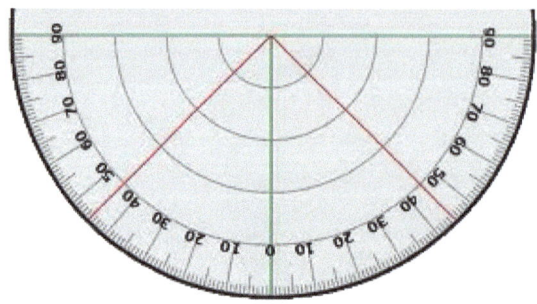

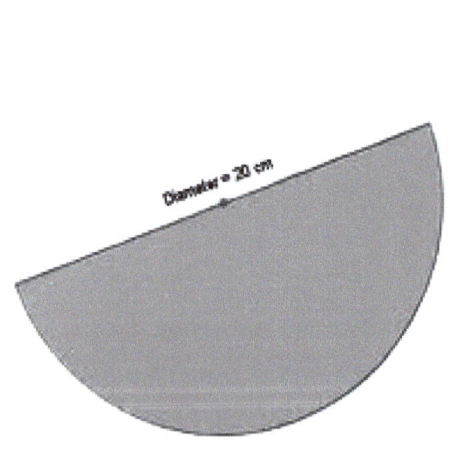

Hemisphere:

Quadrilateral:

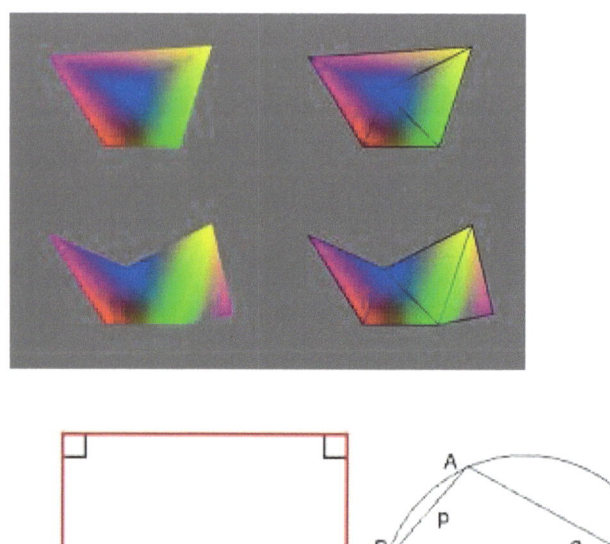

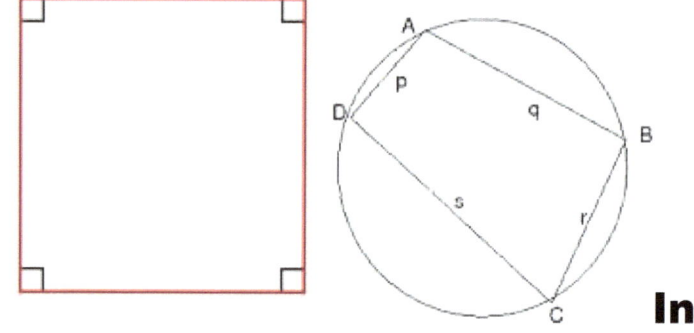

In this lesson we tried to fix the basic concepts of shapes in the tiny mind of your kid.

Hereby I must declare that there are more shapes other than above shapes like oval,octagon,heptgon,hexagon,d-ifferent types of triangle,hyperbole,parabola,oblo-ng etc.

There are so many objects in our home like comb,bucket,bottle,mirror,mobile etc ;while using these objects you must ask to your kid about the shape of that particular object.

Lesson 03 A Point

A point:

This space is filled with infinite number of points .A point is nothing but simply it is a mark made by a pointed object like a needle. .

Space as collection of points

A point

All the object, diagrams, figures, pictures or images are made of points.

Every thing in the space is a collection of points.

A point is named by a capital letter:

. B point B

.C point C

In geometry a point is the most important unit. A Point has neither length nor breadth. As in number system 0(zero) has an important special place. How a cell is the basic unit of life, in the same way we can state that a point is the basic unit of geometry.

_____

Lesson 04 A Line segment

segmentAB

A line segment is a set of fixed number of points .It has a fixed length. It has two end points .It can neither be extended nor be decreased from either end.

Railway track, electrical wires, rulers etc they all have fixed length. We can name the line segment by using capital English alphabets like AB or BA.

Here it is important to note that AB=BA.

(Line) segments are used to draw enclosed geometrical figures that have fixed dimensions (measurement).

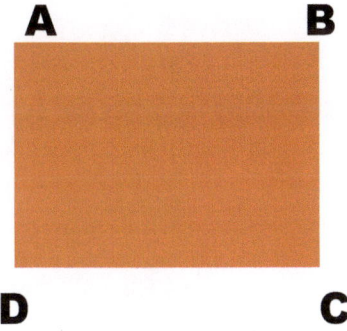

Quadrilateral ABCD is made of 4 segments seg AB, seg BC, seg CD and seg AD.

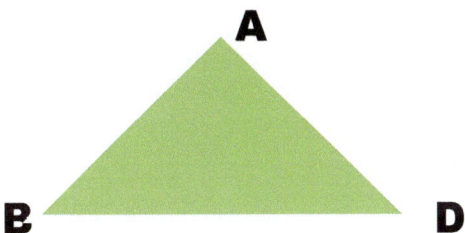

Triangle ABD has 3 line segments namely seg AB, seg AD, and seg BD.

Lesson 05 A Ray

A ray is a collection infinite of number of points which has a beginning but no end.

A ray can be extended in one direction.

A Ray is named as

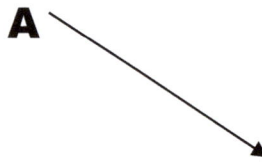

Ray AM; Where Point A is the origin and point M indicates in which direction it goes.

Sunrays,ray digrams,arrows for showing paths etc are the examples where rays are used in our daily

life.

Rays are used to drw angles.

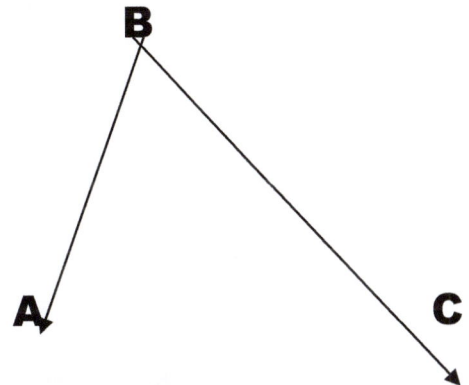

Angle ABC.

In future you will study how rays are important to define an angle.

When three non-collinear rays intersects in 3 distinct points they form a triangle in this case 3 points and 3 rays lie in same Surface(plane).

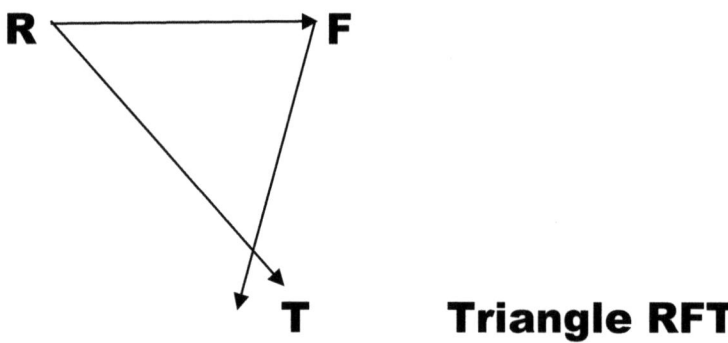

_____

Lesson 06 A Line

A line: <mark>A line is the union of two rays having common origin; and moving in opposite direction.</mark> It has arrows on both the ends that state that a given line is longer than the length it appears. A line has no fixed length.

In other words one can say that a line is the collection (set) of infinite number of points.

Birds flying in opposite direction having no fixed Destination can be an example of a line. A line has one dimension that is length (but indefinite).

Line PQ or Line QP

Points which lie on the same line are called collinear points:

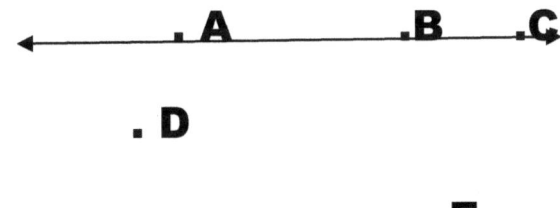

Here points A, B, and C are collinear points where as points D and E are non-collinear points.

Lesson 07 A Plane

A plane: A plane is a set (collection) of infinite number of points that makes a flat surface with no thickness.

Since a plane consists infinite number of points, it can be expanded in all direction.

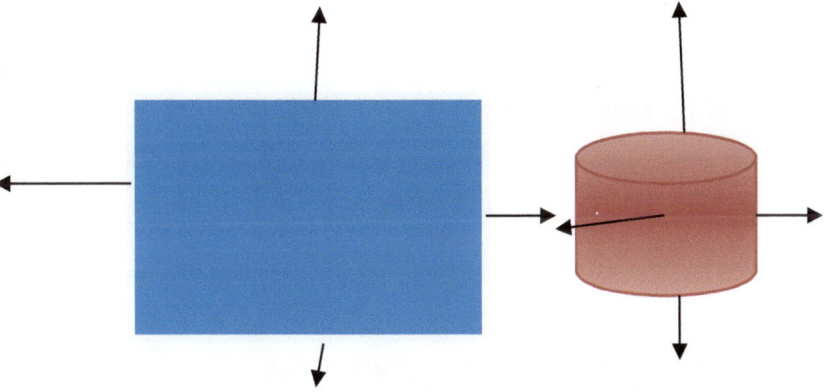

Arrows indicates that a plane is infinitely large surface. A square, a circle, note-book pages etc. are the examples of plane.

Hereby, I must state that there are two types of Plane: 1. two dimensional and 2. Three dimensional.

Detail study of shapes and plane is given in volume 2 of this series.

_____

Let's check your understanding

1. Draw the figures: A ray, a line, a point, a plane.

2. State true or false:

1. A line has fixed length.

2. A ray cannot be measured.

3. An angle can be drawn with 2 rays.

4. A plane is a flat surface.

5. A point has no dimension (length or breadth)

3. Name the following:

1. A geometrical figure has no dimension.

2. A geometrical figure has two end points.

3. A geometrical figure has 3 sides.

4. Name the following figures:

$\xrightarrow{\hspace{3cm}}$

1. A   B

Answer:

2.  P   C

Answer:

3.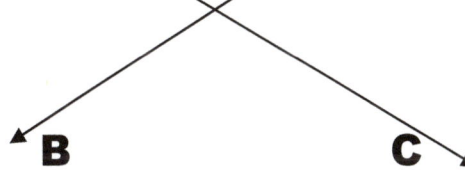

Answer:

4.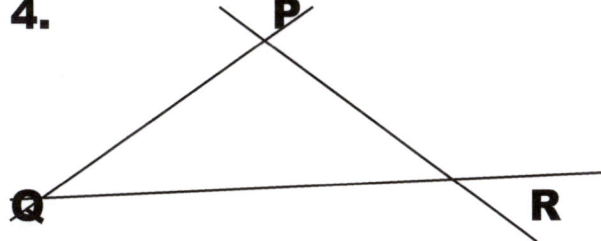

Answer

5.

Answers

Question2

1. false

2. True

3. True

4. True

5. True

QUESTION 3.

1. POINT

2. SEGMENT

3. TRIANGLE

QUESTION 4.

1. Ray AB; 2. Line PC; 3. Angle BAC or Angle CAB

4. Triangle PQR or any 3 letter name of triangle

5. Quadrilateral

Let's see where we use Maths now first you try, think and tell where you don't find Maths?

While shopping we need good skill of counting money.

Profit, loss, commission, expiry date of a product etc all these terms are related with Maths.

Just go through the activities of the daily routine, you will find that this life is ruled by Maths. Today each and everything can be measured, Maths has made it possible.

Maths makes us more accurate.

We cannot imagine life without these things and these things cannot be described without Maths.

These eatables will not have been delicious if cook or chef doesn't have sound knowledge of Maths.

This chair illustrates an obtuse angle.

These vegetables are similar to which geometrical figure?

Aren't they looking like ==spheres?==

Now you must have understood that Maths is everywhere.

What does volume02consist?

In volume 1, we just introduced the basic Mathematical concepts, there are so many Things which remained unexplained will be covered in other volumes.

What does volume 2 consists?

Volume 2 of this series consists detail Study of shapes, one dimensional, two dimensional and three dimensional geometrical figures; it also explains:

How to teach Tables? How to introduce addition and subtraction? How to teach multiplication and division?

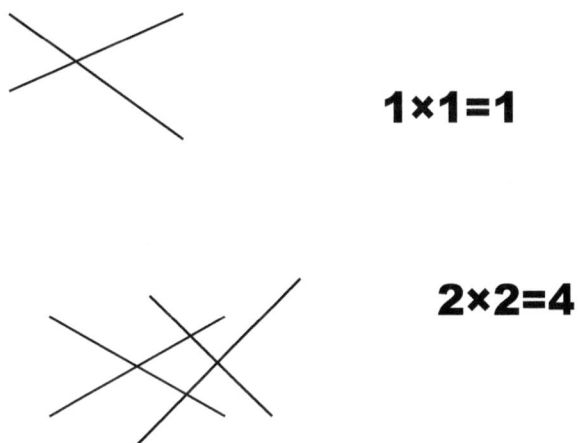

1×1=1

2×2=4

Total how many volumes are there in this series?
In all two volumes are there in this series.

_____

ACKNOWLEDGEMENT

I WOULD LIKE TO THANK WWW.CREATESPACE.COM TEAM FOR PROVIDING THEIR VALUABLE SUPPORT FOR *PUBLISHING*

Let's play with Maths.

TO THE READERS

Dear Readers,

I have taken due efforts to make this book beautiful and errorless, but still there may be some valuable suggestions from your side to improve it further, feel free to ventilate your ideas with me. You can contact me directly through my email id: sunilkushwaha81@gmail.com.

Thank you,

Regards,

Sunil Kushwaha

www.ingramcontent.com/pod-product-compliance
Lightning Source LLC
Chambersburg PA
CBHW051052180526
45172CB00002B/603